AF821971

SOTHIS PRESS

Sothis Press aims to completely reimagine the public domain book with an emphasis on universal access to captivatingly designed print editions. We believe that technology will enable every single book ever published to be perpetually accessible in beautiful print form. Sothis is beauty in ubiquity.

www.sothis.press
1 Waterhouse Square, London, ECIN 2ST

Tight Bounds on the Complexity of the Boyer-Moore Pattern Matching Algorithm

Richard Cole

SOTHIS PRESS

ISBN 978-1-0365-0329-1

Tight Bounds on the Complexity of the Boyer-Moore Pattern Matching Algorithm
is in the public domain.

This edition is published by
Sothis Press The company
is registered in England and Wales.
The Company Number is 14900389.

Tight bounds on the complexity of the Boyer-Moore pattern matching algorithm

Richard Cole

Technical Report 512

June 1990

Tight bounds on the complexity of the Boyer-Moore pattern matching algorithm

Richard Cole

Technical Report 512

June 1990

Tight bounds on the complexity of the Boyer-Moore pattern matching algorithm

Richard Cole*

Courant Institute, New York University

Abstract

The problem of finding all occurrences of a pattern of length m in a text of length n is considered. It is shown that the Boyer-Moore pattern matching algorithm performs roughly $3n$ comparisons and that this bound is tight up to $O(n/m)$; more precisely, an upper bound of $3n - \frac{n}{m}$ comparisons is shown, as is a lower bound of $3n(1-o(1))$ comparisons, as $\frac{n}{m} \to \infty$ and $m \to \infty$. While the upper bound is somewhat involved, its main elements provide a quite simple proof of a $4n$ upper bound for the same algorithm.

1 Introduction

Pattern matching is the problem of finding a pattern of length m in a string of length n; often, all occurences of the pattern are sought. This problem is well studied and is a staple of text books on algorithms (for instance, [AHU73,BB88]). It is an important subproblem in a number of domains including text editing, symbol manipulation and data retrieval.

The best known algorithms for this problem are the Knuth-Morris-Pratt algorithm [KMP77] and the Boyer-Moore algorithm [BM77] (for short, we refer to these as the KMP and BM algorithms, respectively). Both these algorithms are linear time; the bound for the KMP algorithm is very straightforward, the bound for the BM algorithm, considerably less so. An interesting aspect of the BM algorithm is that on average (in probabilistic settings) it takes sublinear time; this effect is observed in practice too. A recent study of this behavior is given in [BGR90].

Both the KMP and BM algorithms begin by computing a shift function (we review the shift function for the BM algorithm later). Following the precomputation of the shift function, the actual match is carried out. The complexity of the algorithm is usually stated in terms of the number of comparisons required for the matching stage (excluding the precomputation stage). The KMP algorithm requires $2n - m$ comparisons in the worst case and this is a tight bound for $m \geq 2$. For the Boyer-Moore algorithm, the first linear bound was given by Knuth in [KMP77], a bound of $7n$ comparisons; this proof is difficult. In 1980, Guibas and Odlyzko gave a less complex proof [GO80], obtaining a bound of $4n$ comparisons; still, their proof, some eight journal pages, is not easy reading. Guibas and Odlyzko also conjectured that $2n$ comparisons might be the correct bound. Our contribution is twofold.

- We give a bound, tight up to lower order terms, of roughly $3n$ comparisons, thereby disproving the just mentioned conjecture.

*The work was supported in part by NSF grants CCR-8902221 and CCR-8906949.

- In addition, the basic elements of this proof provide a direct and straightforward demonstration of an upper bound of $4n$ comparisons. This proof is some three pages long.

The above bounds for the BM algorithm assume that the pattern is not *semi-cyclic*, i.e., of the form wv^k, where w is a proper suffix of v and $k \geq 2$. Galil [Ga79] showed how to modify the BM algorithm so that a linear bound applies in this case too; in fact, using essentially Galil's modification, our bounds apply unchanged to such patterns.

Another approach to improving the bound on the number of comparisons for the Boyer-Moore algorithm is to modify the algorithm so that it no longer necessarily compares characters of the pattern in consecutive right to left order. Such a modification was given in [AG86]; they thereby obtain a bound of $2n - m + 1$ character comparisons, at the cost of a more complex control structure (this algorithm remembers all characters of the text that have been matched and thereby avoids comparing text characters successfully more than once).

Many other types of pattern matching algorithms have been studied; these include matching several strings simultaneously [AC75], real-time matching [Ga81], matching in constant space [GS83,CP89], randomized algorithms [KR87], parallel algorithms [Vi85,Vi90], approximate matching [LV89,GP89,Uk85] two-dimensional matching [ALV90].

The remainder of the paper is organized as follows. In Section 2, we briefly review the Boyer-Moore algorithm. In Section 3, we prove the $4n$ upper bound for non-semi-cyclic patterns; in Section 5, we extend this upper bound to an upper bound of roughly $3n$ comparisons. In Section 4, we show the lower bound of roughly $3n$ comparisons. In Section 6, we extend the results to semi-cyclic patterns. Finally, for the sake of completeness, in the Appendix, we describe how to compute the shift functions.

2 The Boyer-Moore Algorithm: A Review

To find occurences of the pattern in the text, the BM algorithm tests whether given substrings, t', of the text, of length m, match the pattern; each such test is called an *attempted match*. An attempted match is performed as follows: The characters of the pattern are compared, one by one, in right to left order, with the corresponding characters of substring t', until either a mismatch is found or the match is complete. When an attempted match completes (either by a mismatch or by finding a match) the pattern is shifted to the right by the maximum distance consistent with not missing any potential matches. This shift is determined by a shift function, described in the next paragraph; in fact, the actual shift may be smaller than this maximum. Following the shift, another attempted match is performed. This procedure is continued until the pattern is shifted beyond the right end of the text. The initial attempted match is with the leftmost m characters of the text.

The shift is determined by two shift functions. If there is a mismatch, the first shift function, the *occurence shift*, implicitly provides the location of the rightmost character, c, in the pattern, if any, that matches the mismatched character in the text. If c is present, the shift specified by the occurence shift would cause c to become aligned with the mismatched text character; while if c is not present, the shift aligns the leftmost character of the pattern with the text character immediately to the right of the mismatched text character. The second shift function, the *matching shift*, is illustrated in Figure 1; it specifies the smallest shift such that the shifted pattern matches the unshifted pattern on all the characters that were successfully matched, and fails to match the unshifted pattern on the mismatched character; if there is no such shift, then the smallest shift that causes a prefix of the shifted pattern to match a suffix of the matched characters is specified; if there is no shift of this type either, then a shift of length m is specified. It is usual to take the maximum of the occurence and matching shifts as the shift for the Boyer-Moore algorithm. In fact, one could use the matching shift alone. Whichever is done, our upper and lower bounds apply. The presence of the occurence shift is desirable, however, since it helps assure a sublinear behaviour in practice. The only difficulty is that the proof of the upper bound becomes more involved, although the essentials are unchanged.

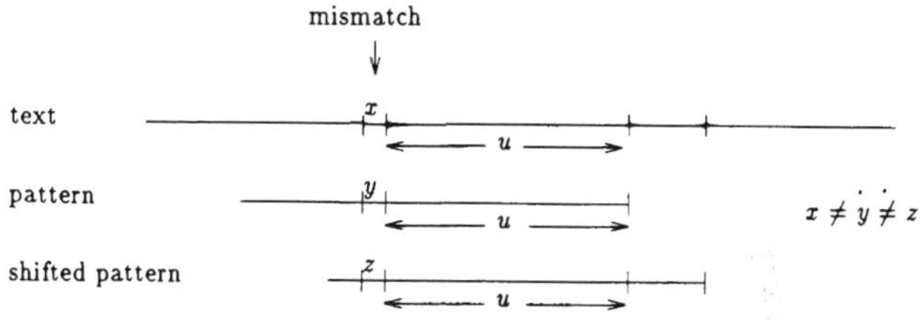

Figure 1: The matching shift

3 A Simple Upper Bound: 4n Comparisons

The analysis uses a simple potential function:

$$3\cdot \text{ \# positions not yet shifted over} + \text{ \# unread text characters}$$

An unread text character is one that has not been involved in a comparison so far. Each shift and the comparisons performed in the associated attempted match are shown to have an amortized cost of at most zero. The initial potential is $4n$ (a value of $4n - 3m$ might be expected, but we have to allow for the final shift, which could move the pattern up to m positions to the right of the text).

Before analyzing the possible shifts, we need a definition and a few results.

Definition. A string u is *semi-cyclic* if it can be written in the form wv^k, where w is a proper suffix of v and $k \geq 2$. If $w = \epsilon$, u is said to be *cyclic*. Also, u is said to be, respectively, *semi-cyclic*, *cyclic* in v. It is convenient to extend this terminology to allow the wording u is semi-cyclic (resp. cyclic) in v to include the case $k = 1$; no ambiguity will result.

Lemma 1 *Let x and y be two non-empty strings. If $xy = yx$ then there is a string z such that both x and y are cyclic in z.*

Proof. The proof is by induction on $|x| + |y|$. If $|x| = |y|$, then take $z = x$ $(= y)$; the result follows. Otherwise, without loss of generality, suppose that $|x| < |y|$. Then x is a prefix of y, so we can write $y = xy_1$. Note that $y_1 \neq \epsilon$. Substituting gives $xxy_1 = xy_1x$; i.e., $xy_1 = y_1x$. The result now follows by induction. •

Corollary 1 *Suppose that v is a proper cyclic shift of w. If $v = w$, v is cyclic.*

Proof. Since v is a proper cyclic shift of w, we can write $w = xy$ and $v = yx$, where $x, y \neq \epsilon$. By Lemma 1, there is a string z with $x = z^i$ and $y = z^j$, for some $i, j \geq 1$. So $v = z^k$, for some $k \geq 2$; i.e., v is cyclic. •

A few more definitions are helpful. $right(v)$ denotes the rightmost character of v. $rightmost(v,$ denotes the rightmost substring of u equal to v. $left(v)$ and $leftmost(v, u)$ are defined analogously.

To begin with, we suppose that only the matching shift is used; subsequently we remove this assumption. Recall that we are assuming the pattern is not semi-cyclic.

Suppose that an unsuccessful attempted match, called the *current attempted match*, causes a shift by distance s. Let u be the suffix of the pattern of length s. Suppose that $u = v^k$, $k \geq 1$, where v is not cyclic. Let t be the portion of text matched in the current attempted match. Clearly, if $|t| \leq 3s$, the current attempted match has amortized cost at most 0 (since the rightmost character of t was unread prior to the current attempted match). So suppose that $|t| > 3s$. Then t must be semi-cyclic in u and hence in v. Our goal is to show that prior to the current attempted match only the following characters of t can have been read: The leftmost $|v| - 1$ characters, and the rightmost $2|v| - 1$, excluding the rightmost character itself (see Figure 2).

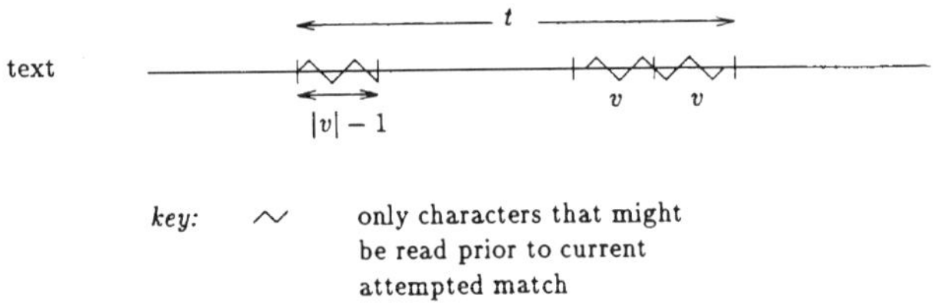

Figure 2: Characters already read

An attempted match AM, which precedes the current attempted match, and begins by comparing a character of t, is called an *early attempted match* (with respect to the current attempted match).

Lemma 2 *Let AM be an early attempted match. Then $right(p)$ is not aligned with $right(v)$ for any substring v in t.*

Proof. See Figure 3. For suppose that the pattern were so aligned. Then a mismatch occurs immediately to the left of $left(t)$ and $right(p)$ would be shifted to distance s to the right of $right(t)$, contradicting the fact that AM is an early attempted match. (For any shorter shift by a multiple of $|v|$ would place the same character in the mismatch location; any other shorter shift causes a proper cyclic shift of v (in p) to be aligned with an instance of v in t, which cannot occur by Corollary 1, since v is not cyclic.) •

Lemma 3 *Let AM be an early attempted match. Then AM performs at most $|v|$ comparisons with characters of t. Further, if there are $|v|$ such comparisons, the last comparison is a mismatch.*

Proof. If $|v|$ characters of t were matched, by Corollary 1, v would be cyclic (since, by Lemma 2, $right(p)$ is not aligned with $right(v)$ for any substring v in t). •

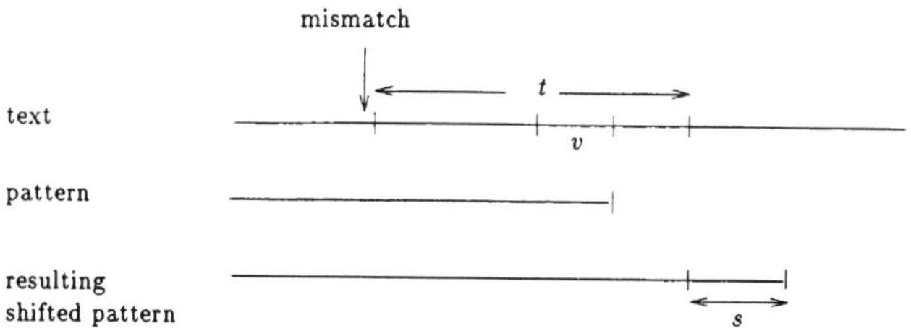

Figure 3: Proof of Lemma 2

Lemma 4 *Let AM be an early attempted match. right(p) is either aligned with a character in rightmost(v, t) or with one of the leftmost* $|v| - 1$ *characters in t.*

Proof. See Figure 4. For if *right(p)* is elsewhere, the first $|v|$ comparisons, if there were that many, would be with characters in t. Then, by Lemma 3, there must be a mismatch on or before the $|v|$th comparison; i.e., there are at most $|v|$ comparisons. Let $\bar{v}$ denote the substring v of t containing the character with which *right(p)* is aligned. But then the shift at most moves *right(p)* to *right*$(\bar{v})$ (for this shift would produce a match with all the text characters compared in the attempted match and so certainly suffices). In fact, if *right(p)* is shifted less far, eventually, following a sequence of such attempted matches, *right(p)* is aligned with *right*$(\bar{v})$. But this contradicts Lemma 2. •

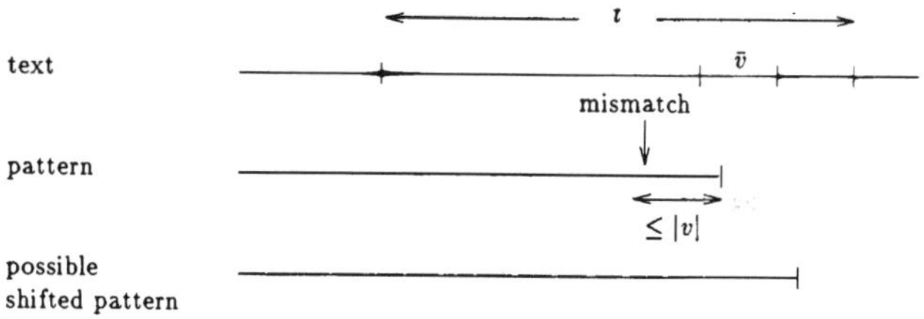

Figure 4: Proof of Lemma 4

We now conclude:

Lemma 5 *Prior to the current attempted match, at most* $3|v| - 3$ *characters of t have been read.*

Proof. See Figure 2. By Lemma 4, in the early attempted matches, *right(p)* is either aligned with one of the rightmost $|v|$ characters of t or with one of the leftmost $|v| - 1$ characters of t. By Lemma 3, each of the former attempted matches performs at most $|v|$ comparisons. Since *right(t)* is unread prior to the current attemped match, the lemma follows. •

We can now bound the amortized cost of the current attemped match.

Lemma 6 *The current attempted match has amortized cost at most zero.*

Proof. Let s be the distance shifted. The number of comparisons performed is $|t|+1$. If $3s \geq |t|$, then the potential is reduced by at least $3s+1 \geq |t|+1$ (since the rightmost character matched had been unread); the result follows in this case.

So suppose that $3s < |t|$. Then by Lemma 5, the number of characters of t read is at most $3|v|-3 \leq 3s-3$. So the reduction in potential is at least $3s+(|t|-3s+3)=|t|+3$. The result follows in this case also. •

It remains to consider the case that an attempted match matches the complete pattern. By assumption, the pattern is not semi-cyclic; so the shift must be of length at least $\frac{m+1}{2}$; it follows that this attempted match also has amortized cost at most zero.

We have shown:

Theorem 1 *If the Boyer-Moore pattern matching algorithm determine its shifts using only the matching shift rule, then it performs at most $4n$ comparisons when matching a non-semi-cyclic pattern of length m against a text of length n.*

Now, we extend the result to allow the use of both the occurrence and matching shifts in the case of non-semi-cyclic patterns. Consider the proof of Lemma 6. We redefine s to be the length of the shift determined by the matching shift (which is either equal to or smaller than the length of the actual shift). In order for Lemma 6 to continue to hold, it suffices that Lemma 5 remain true in the presence of the occurrence shift. But it is a simple matter to check that Lemmas 2-5 continue to hold in the presence of the occurrence shift. Thus:

Theorem 2 *The Boyer-Moore pattern matching algorithm performs at most $4n$ comparisons when matching a non-semi-cyclic pattern of length m against a text of length n.*

4 The Lower Bound

We give example patterns, p, of length $m=2k-1$, and texts t, which demonstrate the lower bound of $3n(1-o(1))$ comparisons (as $\frac{n}{m} \to \infty$ and $m \to \infty$).

The basic idea is to have frequent shifts by k on attempted matches comprising $2k-1$ comparisons. In addition, in order to have roughly 3 comparisons per text character, we also seek to have short shifts on attempted matches of roughly k comparisons.

In particular, we choose $p=a^{k-1}ba^{k-1}$ and $t=a^{k-1}(aba^{k-1})^{\lambda}$. Let $v'=aba^{k-1}$.

Consider a situation in which the b in the pattern is aligned with the left end of a v' in the text, denoted $\bar{v}'$. (See Figure 5.) The attempted match will be of length $k-1$ and it will cause a shift of length 1. The next attempted match will have $right(p)$ aligned with $right(\bar{v}')$. The attempted match will be of length $2k-1$ and it will result in a shift of length k. We return to the first situation.

We note that t has been chosen so that the initial situation is the first situation. We have shown:

Theorem 3 *There exist patterns of length $m=2k-1$ and texts of length $n=\lambda(k+1)+(k-1)$, for any integers $k \geq 2$ and $\lambda \geq 1$, for which the Boyer-Moore pattern matching algorithm performs $\frac{3k-2}{k+1}(n-k+1)=(n-\frac{m-1}{2})(3-\frac{10}{m+3})$ comparisons. This is $3n(1-o(1))$ comparisons as $\frac{n}{m} \to \infty$ and $m \to \infty$.*

For even length patterns, with $p=a^{k-2}ba^{k-1}$, for $k \geq 2$, and t one character shorter at the left end, we obtain a similar bound of $\frac{3k-3}{k+1}(n-k+2)=(n-\frac{m-2}{2})(3-\frac{12}{m+4})$ comparisons.

text	⋯ a b a a | a b a a | a b a a ⋯	
pattern	a b *a c*	mismatch at *a*
first shift	a b a *a*	match
second shift	a b a	iterate

Figure 5: The lower bound for $k = 3$

5 An Upper Bound of Roughly 3n Comparisons

A more careful analysis, in the style of Section 3, yields an upper bound of slightly fewer than $3n$ comparisons. We use the following potential function:

$$2 \cdot \#\text{ positions not yet shifted over} + \#\text{ unmarked characters}.$$

We show that each attempted match has an amortized cost of at most -1. Initially, all the characters of the text are unmarked. The unmarked characters are treated in the same way as unread characters in Section 3. When an unmarked character is read it becomes marked. However, a marked character can be unmarked anew, as follows: Whenever an attempted match would have an amortized cost less than -2, the leftmost characters compared in the current attempted match are unmarked so as to increase the amortized cost to -2 (or until all the characters read in the current attempted match are unmarked, whichever occurs sooner); actually, as we see below, there is a special case for shifts of length 1.

As in Section 3, we begin by considering the variant of the Boyer-Moore algorithm in which only the matching shift rule is used. Also, we are still assuming that the pattern is not semi-cyclic.

As in Section 3, we consider a current attempted match which matches a substring t of the text, and mismatches immediately to the left of $left(t)$. We also consider earlier attempted matches in which $right(p)$ is aligned with a character of t. Let s be the distance shifted in the current attempted match. There are a number of cases to consider (in all the cases apart from Case 1 and Case 3.3.1 we show an amortized cost of at most -2 for the current attempted match, and for Case 3.3.1 this bound can be achieved by a more careful and longer argument).

Case 1. $s = 1$. Let t' be the text read by the current attempted match. Then $t' = ba^j$ for some $j \geq 0$, where $a \neq b$. We claim that the characters of t' had not been read previously. This would give an overall amortized cost of -2 for the attempted match, except that we unmark the rightmost character of t', giving an amortized cost of -1. It remains to demonstrate the claim. If there had been an earlier attempted match with $right(p)$ aligned with some character of t', then the mismatch would have been at character b of t', and the resulting shift would have moved $right(p)$ beyond $right(t)$; so there was no such earlier attempted match; i.e., all of t' was unmarked immediately prior to the current attempted match.

Remark 1. Henceforth, we can assume that at each attempted match the rightmost two characters compared (if at least two are compared) are unmarked.

Case 2. $|t| < 2s$. The decrease in potential is at least $2s+2$. The cost of the current attempted match is $|t| + 1$. So the amortized cost of the current attempted match is at most -2.

Case 3. $|t| \geq 2s$ and $s > 1$. Then $t = wv^k$, for some $k \geq 2$, where w is a proper suffix of v, v is non-cyclic, and s is an integer multiple of $|v|$. Lemmas 2-4 continue to hold. A little more notation is helpful. Let w_L, v_L, v_{R1}, v_{R2} denote, respectively, $leftmost(w,t)$, $leftmost(v,t)$, $rightmost(v,t)$, and the second rightmost v in t. See Figure 6.

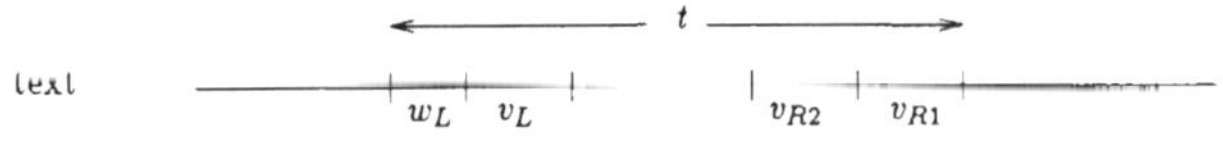

Figure 6: Further Notation

Case 3.0. There is no early attempted match. Then the current attempted match has amortized cost at most $-2s+1 \leq -3 \leq -2$.
Case 3.1. AM, the first early attempted match, has $right(p)$ aligned with a character of v_{R1} other than $right(v_{R1})$. Then at most $2|v|-3$ characters of t will have been read and remain marked prior to the current attempted match (for among the characters of v_{R1} the rightmost two are unmarked, and by Lemma 3, besides characters of v_{R1}, only the $|v|-1$ rightmost characters of v_{R2} can have been read). So the amortized cost of the current match is at most -2.

Before considering the next case, we prove a lemma.

Lemma 7 *Consider an early attempted match, AM_1, which results in a shift of length s_1.*

(i) *If AM_1 matches at least $|v|$ characters, then s_1 is an integer multiple of $|v|$. Also, the number of characters matched by AM_1 is exactly $|t|+s-s_1$.*

(ii) *While if AM_1 matches fewer than $|v|$ characters, then $s_1 < |v|$.*

Proof. First note that the shift does not move $right(p)$ to the right of $right(t)$ (since this is an early attempted match). Thus $s_1 < |wv^k|$.

Now, we prove Part (i). Let $\bar{v}$ be the string in the text which matches $rightmost(v,p)$ during attempted match AM_1. Following the shift, the substring $\bar{\bar{v}}$ of p aligned with $\bar{v}$ must be the pattern v; further $\bar{\bar{v}}$ is part of the suffix of p of length $|wv^{k+1}|$ (as $s_1 < |wv^k|$), which suffix is semi-cyclic in v; thus $\bar{\bar{v}}$ is a cyclic shift of v; by Corollary 1, this cyclic shift must be the trivial shift; i.e., s_1 is an integer multiple of $|v|$.

Suppose that the number of characters matched is greater than $|t|+s-s_1$. Let c be the $(|t|+s-s_1+1)$th character of the text matched by AM_1. Then c will not match the character of p with which it is aligned following the shift by distance s_1 (for this is the $(|t|+s+1)$th rightmost character of p, which differs from characters $l|v|$ to its right in p, for any $l \geq 1$). But this contradicts the definition of the matching shift. So suppose that the number of characters matched is fewer than $|t|+s-s_1$. Then consider the character, c, of the text with which AM_1 mismatches; the character of p aligned with c following the shift by s_1 is the same as the character of p aligned with c before the shift. Again, this contradicts the definition of the matching shift. Thus exactly $|t|+s-s_1$ characters are matched by AM_1.

We turn to Part (ii). Suppose that $s_1 > |v|$. Suppose c characters are compared by AM_1. See Figure 7. Then a shift by $s_1-|v|$ would also be a legal shift, in that the characters of the pattern aligned with the c text characters compared by AM_1 are identical in a shift by s_1 and a shift by $s_1-|v|$ (this follows from the fact that the portion of p semi-cyclic in v is of length at least $s_1+|v|$). So $s_1 \leq |v|$. But a shift by $|v|$ is not possible for this would replace the mismatched character of the pattern by the same character, which is not a legal shift. So $s_1 < |v|$. •

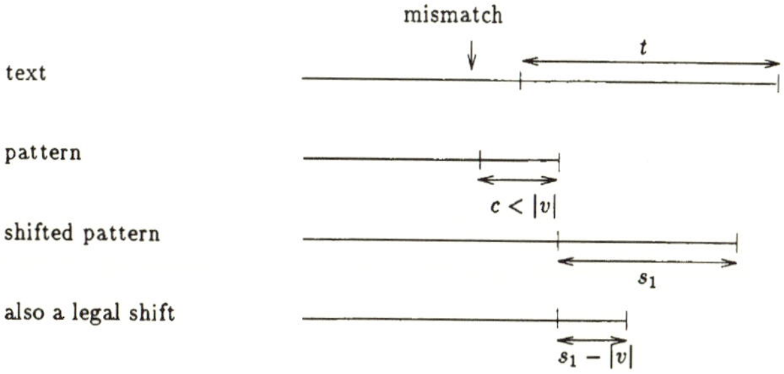

Figure 7: Proof of Lemma 7

Case 3.2. AM_1 is an early attempted match in which $right(p)$ is aligned with a character in w_L, which causes $right(p)$ to be shifted distance s_1 so that it is aligned with a character in v_{R1}. Note that $s_1 > |v|$ (for $right(p)$ shifts over the whole of v_L). By Lemma 7, the shift has length an integer multiple of v. Suppose that at the start of AM_1 $right(p)$ is aligned with the rth rightmost character of w_L. Then, following AM_1, $right(p)$ is aligned with the rth rightmost character of v_{R1}. By Lemma 3, at most $|v| - 1$ characters of the text immediately to the left of this location can be read prior to the current attempted match. So the number of characters of t that can be marked prior to the current attempted match is bounded as follows: At most $|w| - r + 1$ characters of w_L, plus at most $\max\{(r-2) + (|v|-1), 0\} = (r-2) + (|v|-1)$ characters at the right end of t. This is a total of $|w| + |v| - 2 \leq 2|v| - 3$ characters. Hence the current attempted match has amortized cost at most -2.

Case 3.3. AM_1 is an early attempted match in which $right(p)$ is aligned with a character in v_L, which causes a shift, by distance s_1, that aligns $right(p)$ with a character in v_{R1}. By the argument of the proof of Lemma 4, the characters matched by AM_1 include at least all the characters up to the $left(t)$ (for if not the resulting shift would leave $right(p)$ aligned with a character of v_L). Let x be the suffix of p that overlaps v_L prior to the attempted match AM_1, and let y be the suffix that overlaps v_{R1} after this attempted match. We consider two subcases depending on the number of characters matched by AM_1.

Case 3.3.1. At least $|v|$ characters are matched. See Figure 8. By Lemma 7, s_1 is an integer multiple of $|v|$. So $x = y$. Consider the next attempted match, AM_2, also an early attempted match. Its mismatch occurs at the $(|wx| + 1)$th character compared (for the $(|wx| + 1)$th characters of the text compared by AM_1 and AM_2 are not equal, and as $|wx| < |v|$ by Lemma 3 applied to AM_1, this is a character matched by AM_1). Suppose AM_2 results in a shift by s_2. Let the suffix of p of length s_2 be z.

Suppose that $|z| \leq |w|$. Then $wx = z'z^l$, where z' is a proper suffix of z and $l \geq 1$. Let $\bar{z}$ be the shortest non-cyclic string such that z is cyclic in $\bar{z}$ (possibly $z = \bar{z}$). x is cyclic in $\bar{z}$ (for consider $\bar{\bar{z}} = rightmost(\bar{z})$ in v_{R2}, and consider the substring, $\bar{\bar{\bar{z}}}$ of p aligned with $\bar{\bar{z}}$ following the shift of AM_1; if $x = y$ is not cyclic in $\bar{z}$, then $\bar{\bar{\bar{z}}}$ is a proper cyclic shift of $\bar{z}$, which, by Corollary 1, contradicts the minimality of $\bar{z}$). But the fact that the shift due to AM_2 causes a match on the characters matched by AM_2, means that p has a suffix, semi-cyclic in $\bar{z}$, of length $|wxz|$. Since v is also a suffix of p and $|w| + 1 \leq |v|$, the rightmost $|w| + 1$ characters of v

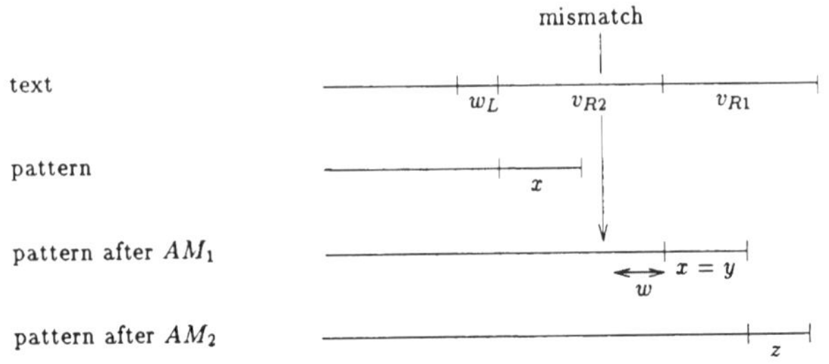

Figure 8: Case 3.3.1

are semi-cyclic in $\bar{z}$; so there is no mismatch by AM_2 at the location assumed (the $(|wx|+1)$th character compared by AM_2). This is a contradiction.

So $|z| > |w|$. Then, prior to unmarking read nodes, the amortized cost of AM_2 would be $-2|z|$ since all the characters compared were previously unread (as $|w| + |x| < |v|$). So $\min\{2|z| - 2, |wx| + 1\} \geq |w|$ characters are unmarked. Hence, following attempted match AM_2, the $|w|$ characters immediately to the right of $leftmost(x, v_{R2})$ are unmarked; by Lemma 3, these characters are not read anew prior to the current attempted match. So the number of marked characters in t at the time of the current attempted match is bounded by: $(|v| - 2)$ characters in v_{R1}, $(|v| - |w|)$ characters between v_{R2} and v_L (note that v_L is not necessarily equal to v_{R2}) and $|w|$ characters in w_L; this is a total of at most $2|v| - 2$ marked characters. Hence the amortized cost of the current attempted match is at most -1.

Case 3.3.2. Fewer than $|v|$ characters are matched. Let r be the number of characters matched. Note that $r > |w|$.

First, suppose that $s_1 > r$. Following the shift due to AM_1, at least s_1 unmarked characters can be created; so, in fact, all the characters read by AM_1 are unmarked. These unmarked characters include w_L plus the character immediately to the right of w_L. By Lemma 3, these $|w| + 1$ characters are not reread until the current attempted match. So there are at most $2|v| - 3$ marked characters in t at the time of the current attempted match, and hence the amortized cost of the current attempted match is at most -2.

Second, suppose that $s_1 \leq r$. We show this case cannot arise. So suppose this case occurs. As $s_1 < |v|$, $|y| < |x|$. See Figure 9. Since a suffix of p, of length $|vt|$, is semi-cyclic in v, and the shifted pattern matches the text characters matched by AM_1, y must be a prefix of x. Next, we show that x is semi-cyclic in y. The portion, $\bar{x}$, of the shifted pattern, aligned with the suffix x of p prior to the shift, must match the suffix x. Note that $\bar{x} = yv'$ where v' is a prefix of v. Prior to the shift, $left(x)$ is aligned with $left(v_L)$, so v' is a prefix of x also. It follows that $x = y^i y'$, for some $i \geq 1$, where y' is a proper prefix of y. As y is also a suffix of x (since they are both suffixes of p), using Corollary 1, we conclude that there is a non-cyclic string z such that both x and y are cyclic in z. We also conclude that the suffix of p of length r is of the form $z'z^i$ for some $i \geq 1$, where z' is a proper suffix of z. (For consider the portion of the pattern, following AM_1, which is aligned with the rightmost r characters of the pattern prior to AM_1; it comprises the prefix of v of length $|x| - |y|$, preceded on the left by the suffix of v of length $r - (|x| - |y|)$. Also, it matches the suffix of v of length r. As the prefix of v of length $|x| - |y|$ is cyclic in z the assertion follows.) Also, note that x is a prefix of v (consider the attempted match AM_1 and the portion of t with which x is matched). To avoid v being cyclic in z, we need that in $\bar{v} = rightmost(v,p)$, the suffix of length r overlap with $leftmost(x, \bar{v})$ by fewer than $|z|$ characters; i.e., $r + |x| < |v| + |z|$, so $r < |v| + |y| - |x| = s_1$; this is a contradiction. So this case does not arise.

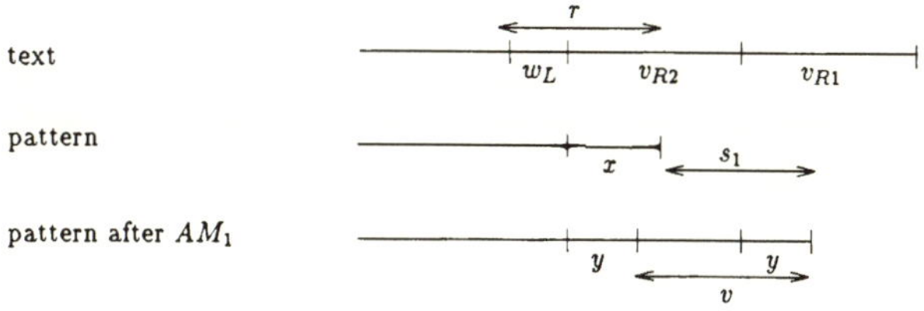

Figure 9: Case 3.3.2

It remains to consider the case that an attempted match matches the complete pattern. As in Section 3, since the pattern is not semi-cyclic, the shift must be of length at least $\frac{m+1}{2}$, and so such an attempted match has amortized cost at most $-3 \leq -2$.

We can conclude:

Theorem 4 *The Boyer-Moore pattern matching algorithm, restricted to using only the matching shift, performs at most $3n - \frac{n}{m}$ comparisons when matching a non-semi-cyclic pattern of length m againt a text of length n.*

Proof. Let r ($\leq m$) be the distance finally shifted beyond the right end of the text. As each shift has amortized cost at most -1, and as each shift is by distance at most m, the amortized cost of the shifts in total is at most $-\frac{n-(m-r)}{m}$. Since the potential drops by at most $3n - 2(m - r)$, the result follows. •

Improving the bound of Case 3.3.1 to an amortized cost of at most -2, improves the bound of the theorem to $3n - \frac{2n}{m}$ comparisons. Actually, this is still too large a bound for it assumes each shift is by distance m (which would result in fewer comparisons being made). Proving a tight bound appears to involve a quite elaborate (and tedious) case analysis, which (in my opinion) is not of sufficient interest to merit being detailed.

It remains to extend the result to incorporate the occurence shift. This is not quite as simple as in Section 3. We follow the analysis used earlier in this section and as in Section 3, s will denote the length of the matching shift. Cases 1, 2, 3.0 and 3.1 apply unchanged. Cases 3.2 and 3.3 require a new analysis. We introduce a new rule for unmarking nodes: When the occurence shift is used, suppose that the length of the occurence shift exceeds that of the matching shift by h; then the rightmost $\min\{2h, |t| + 1\}$ characters compared are unmarked.

Case 3.2. See Figure 10. A new case arises when AM_1 uses the occurence shift. Let x be the suffix of w_L to the right of $right(p)$ prior to attempted match AM_1. Consider the first early attempted match, AM_2, subsequent to AM_1, if any, which compares characters of v_{R2} beyond the rightmost $|x|$ characters. If AM_2 does not exist, then at most $(|v| - 2) + |x| + \max\{0, |w| - |x| - 2\} \leq 2|v| - 4$ characters of t will be marked prior to the current match (recall the at least two characters unmarked due to the occurence shift); while if AM_2 has $right(p)$ to the right of the leftmost $|w| - |x|$ characters in v_{R1}, then at most $[|v| - (|w| - |x|) - 2] + (|v| - 1) + \max\{0, |w| - |x| - 2\} \leq 2|v| - 4$ characters of t will be marked prior to the current match. In these cases the current match has amortized cost at most $-3 \leq -2$.

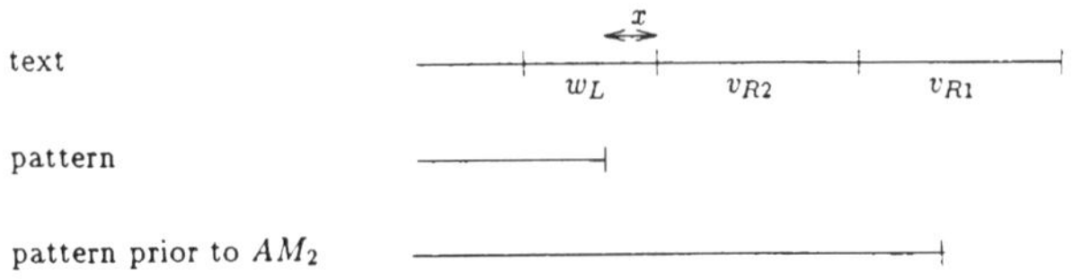

Figure 10: Case 3.2 - the Heuristic Shift

So suppose that AM_2 has $right(p)$ aligned with one of the leftmost $|w| - |x|$ characters in v_{R1}. We first consider the case that $x \neq \epsilon$. Let $\bar{x}$ be the shortest non-cyclic string such that x is cyclic in $\bar{x}$ (possibly $x = \bar{x}$). Since AM_1 matches every character of w_L to the left of x, w is semi-cyclic in x and thus in $\bar{x}$. Let v_x be the longest suffix of v such that $v_x = \bar{x}'\bar{x}^j$ for some $j \geq 1$, where $\bar{x}'$ is a proper suffix of $\bar{x}$. As AM_2 matches the rightmost $\bar{x}$ characters of v_{R2}, and the corresponding matched characters of p are among p's righmost $|w|$ characters, it must be that the substrings $\bar{x}$ in v_{R2} and p are aligned (or else Corollary 1 is contradicted); so $j \geq 2$ (for $rightmost(x, v_{R2})$ is matched, as is at least one string $\bar{x}$ to its right). It follows that exactly the suffix v_x of p is matched. So the matching shift for AM_2 has length greater than $|\bar{x}'\bar{x}^{j-1}|$. The reduction in potential is at least the length of the suffix of v_{R2} compared by AM_2 minus $|x|$ (previously unread text), plus twice the shift ($\geq 2|\bar{x}'\bar{x}^{j-1}| + 2$); this suffices to pay for all the characters read plus the unmarking, in v_{R2}, of all but the rightmost $|x|$ characters. Also, following the shift by AM_2, $right(p)$ is to the right of the leftmost $|w| - |x|$ characters in v_{R1}. Now, by the argument of the last two sentences of the previous paragraph, it follows that the current attempted match has amortized cost at most -2.

Now suppose that $x = \epsilon$. We show that this case cannot occur. For if $x = \epsilon$, the mismatch in AM_1 would occur immediately to the left of $left(t)$. As the heuristic shift moves the rightmost occurence of a character to the mismatch location, since the character placed in the mismatch location is from $rightmost(t, p)$ it must be from $rightmost(v, p)$; but this contradicts the fact that the actual shift was of length greater than $|v|$.

Case 3.3.1. If AM_1 uses the matching shift we argue as in the previous Case 3.3.1 (defining s_2 to be the shift given by the matching shift). So suppose that AM_1 uses the occurence shift. Let x denote the overlap of p and v_L prior to attempted match AM_1. Let $\bar{x}$ be the shortest non-cyclic string in which x is cyclic. Clearly $w = \bar{x}'\bar{x}^i$, for some $i \geq 0$, where $\bar{x}'$ is a proper suffix of $\bar{x}$. Let v_x denote the longest suffix of v semi-cyclic in $\bar{x}$; $|v_x| \geq |wx|$. There are a number of cases to consider.

First, suppose that the matching shift of AM_1 would move $right(p)$ to be aligned with a character of v_{R1}. Since the matching shift, in this case, has length a multiple of $|v|$ (by Lemma 7), following the matching shift, p would overlap v_{R1} by $|x|$ characters. The occurence shift must result in a greater overlap (as it is the shift used). Consider the first subsequent early attempted match, AM_2, that compares a character of v_{R2}, if any. If AM_2 does not exist, or if, in AM_2, $right(p)$ is to the right of the leftmost $|wx|$ characters of v_{R1}, then the number of marked characters in t at the time of the current attempted match is at most

$\max\{|v|-2, (|v|-2-|wx|)+(|v|-1)\}+|w|+|x| \leq 2|v|-3$, and the amortized cost of the current attempted match is at most -2. So suppose that *right(p)* is aligned with one of the leftmost $|wx|$ characters in v_{R1}. There are two subcases to consider, depending on the length of x.

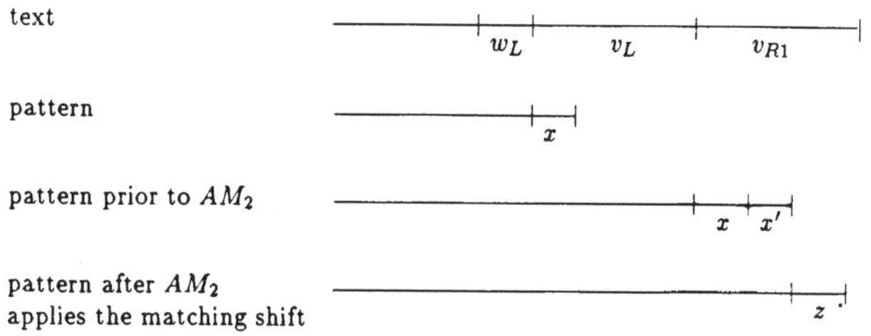

Figure 11: Case 3.3.1 – the Heuristic Shift

Suppose that $|x| \leq |v|/2$. See Figure 11. Let s_2 be the length of the matching shift for AM_2 and let z be the suffix of p of length s_2. Let x' be the overlap of p beyond $leftmost(x, v_{R1})$ at the start of attempted match AM_2; by assumption, $|x'| \leq |w|$. Let $\bar{x}'$ be the shortest non-cyclic string with x' cyclic in $\bar{x}'$. x is cyclic in $\bar{x}'$ (this can be seen by an argument similar to that used in the previous Case 3.3.1 to show x is cyclic in $\bar{z}$). Thus w is semi-cyclic in $\bar{x}'$; we conclude that AM_2 mismatches at the $(|wx|+1)$th character compared (for the longest suffix of v semi-cyclic in $\bar{x}'$ is of length $|wx|$, as AM_1 does not mismatch at the $(|wx|+1)$th character compared). By another argument similar to that of the previous Case 3.3.1, we conclude that $|w| < |z|$. We want at least $|w|+1$ characters of v_{R2} to remain unmarked aside from $leftmost(x, v_{R2})$, for then at most $2|v|-3$ characters of t will be marked at the time of the current attempted match. Then, following the shift due to AM_2, the potential at hand for unmarking characters anew is at least $2|z|-|x|-|x'|$; to obtain $|w|+1$ unmarked characters in v_{R2} aside from $leftmost(x, v_{R2})$, we need

$$(|v|-|x|)-(|w|-|x'|+1)+(2|z|-|x|-|x'|) \geq |w|+1$$

i.e., $|v|-2|x|+2|z| \geq 2|w|+2$, which is true when $|x| \leq |v|/2$.

So suppose that $x > |v|/2$. Then both v and x are semi-cyclic in some non-cyclic y, where $|xy| \leq |v|$ (since x matches a prefix of v). Any early attempted match, subsequent to AM_1, which matches a character of v_{R2} will have matched a suffix of p of length greater than $|x| (> |y|)$ and at least $|x|$ characters of v_{R1}. Also, such an early attempted match can compare at most $|y|$ characters of v_{R2} (for v is not cyclic in y, by assumption, and as at least one character of v_{R1} is compared, the "y" substrings in p and v_{R2} are not aligned). The resulting shift will be of length a multiple of $|y|$, as y is non-cyclic and v is semi-cyclic in y. Furthermore, the resulting shift must align *right(p)* and *right(t)* (as this is the least shift by a multiple of $|y|$ that places a different character in the mismatch location). So the number of marked characters for the current attempted match is at most $|v|+(|w|+|x|-2) \leq 2|v|-3$ (the $|v|$ term is for the characters compared in the attempted match immediately prior to the current attempted match, and the other term is for the characters compared by AM_1; the -2 term arises because at least two characters are unmarked due to the occurence shift).

Second, suppose that the matching shift of AM_1 does not align $right(p)$ with a character of v_{R1}. Then, if $x \leq |v|/2$, the length of the occurence shift exceeds that of the matching shift by at least $|v|/2$; the extra (at least) $|v|$ decrease in potential is used to mark $leftmost(wx, t)$, which then results in at most $2|v| - 3$ characters being marked for the current attempted match. So suppose that $|x| > |v|/2$. Then x and v are semi-cyclic in some non-cyclic y, where $|xy| \leq |v|$. Hence any subsequent early attempted match which compares characters of v_{R2} can compare at most y such characters (for v is not cyclic in y, by assumption, and as at least one character of v_{R1} is compared, the "y" substrings in p and v_{R2} are not aligned). So none of the characters of $leftmost(x, v_{R2})$ are read anew prior to the current attempted match. The occurence shift exceeded the length of the matching shift by at least $|y| + 1$, hence at least the $|y| + 1$ rightmost characters matched by AM_1 are unmarked. Further, as just noted, these characters are not read anew prior to the current attempted match. So the number of marked characters at the time of the current attempted match is at most

$$(|v| - 2) + |y| + (|w| + |x|) - (|y| + 1) \leq 2|v| - 3$$

We conclude that the current attempted match has amortized cost at most -2.

Case 3.3.2. We use the notation from the earlier Case 3.3.2. Again, we are only concerned with the case in which AM_1 uses the occurence shift. The case $s_1 > r$ is handled as before. So suppose that $s_1 \leq r$. Consider the character, c, of t which was mismatched by attempted match AM_1. The character, c', of p that matches c following the shift of AM_1 must be a character of the second rightmost v of p (for the shift was by distance $< |v|$ and fewer than $|v|$ characters were matched, so c′ must be among the rightmost $2|v|$ characters of p; yet as c is to the left of t, c′ cannot be among the rightmost $|v|$ characters of p, for p overlaps t by more than $|v|$ characters following the shift by AM_1). But then this is not the rightmost occurence of this character in p, and so this shift is not the shift given by the occurence shift. This is a contradiction. Hence the case $s_1 \leq r$ does not occur.

We have shown:

Theorem 5 *The Boyer-Moore pattern matching algorithm performs* $3n - \frac{n}{m}$ *comparisons when matching a non-semi-cyclic pattern of length* m *againt a text of length* n.

6 Semi Cyclic Patterns

Now we handle the case of semi-cyclic patterns. So suppose that the pattern, p, is of the form $p = wv^i$, where v is not cyclic, w is a proper suffix of v, possibly empty, $i \geq 2$, and v is the shortest such string.

We use the pattern $p' = uv$, where u is the suffix of v of length $|v| - 1$. Note that p' is not semi-cyclic. In addition, we record the following index j, $0 \leq j < i$. Let $q_j = wv^j$. See Figure 12. We record the largest j such that the text contains an instance $\bar{q}_j$ of the pattern q_j, where $right(\bar{q}_j)$ is aligned with $right(u)$.

Following an attempted match, j is modified as follows. If p' is matched, then the shift is by distance $|v|$ (by Corollary 1, a shorter shift contradicts the fact that v is not cyclic). If $j = i - 1$, then a match of p has been found; j is unchanged. If $j < i - 1$, then j is incremented. If p' is not matched, but at least wv is matched, then j is set to 1 (note that the shift is by distance $s = |v|$ in this case too); while if not even wv is matched, then j is set to 0.

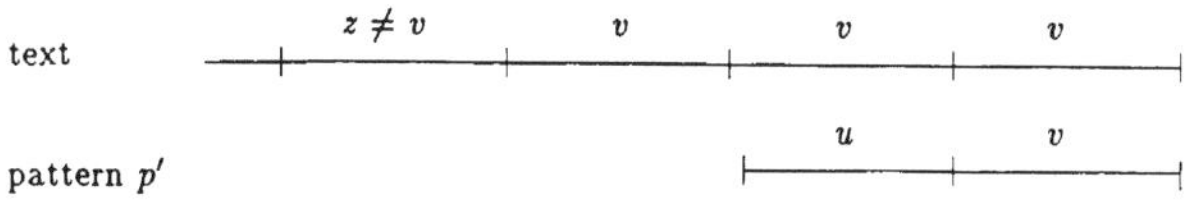

If z has suffix w, $j = 2$
otherwise, $j = 1$

Figure 12: Semi-cyclic patterns

The comparison count for the non-semi-cyclic case applies here too, since p' is not semi-cyclic.

We conclude:

Theorem 6 *The Boyer-Moore pattern matching algorithm performs at most* $3n - \frac{n}{m}$ *comparisons in matching a pattern of length* m *against a text of length* n.

Acknowledgements

I would like to thank Rajamani Sundar for his comments.

A Computing the shift functions

We describe how to compute the shift functions. Descriptions of these procedures have been given previously, for instance in [KMP77,Ry80]. The following notation will be helpful: $p[i,j]$ will denote the substring of the pattern extending from the ith to jth characters inclusive and $p(i)$ will denote the ith character of the pattern.

The occurence shift is computed by a left to right sweep accross the pattern. Suppose the pattern is indexed from 1 to m, from left to right. Then, for each i, $1 \leq i \leq m$, we perform the assignment $o\_shift(p(i)) := i$. This stores, in $o\_shift(c)$, the rightmost occurence of character c in the pattern. If there is a mismatch at $p(j)$, and the mismatched text character is c, the occurence shift is given by $j - o\_shift(c)$. To handle the possibility that c does not occur in the pattern, $o\_shift(c)$ should be initialized to 0, prior to the sweep.

The computation of the matching shift is less simple. A program is given in Figure 13. We introduce an auxiliary shift function, called the kmp_shift. It is the shift function used in some versions of the Knuth-Morris-Pratt algorithm (except that here, we are matching from the right end of the pattern). Its definition follows. Define $l(j)$ to be the smallest index with $l(j) > j$, such that $p[l(j), m]$ is a prefix of $p[j, m]$; if $l(j)$ is not thereby defined, it is set to $m+1$. Then $kmp\_shift(j) = l(j) - j$. It is straightforward to check that kmp_shift is correctly computed in Stage 1.

Next, we explain why the matching shift, $m\_shift(j)$, for $1 \leq j \leq m$, is correctly computed. There are two possibilities.

Either we need to determine a largest index $k < j$ such that $p(j) \neq p(k)$ and $p[j+1, m]$ is a prefix of $p[k+1, m]$ (which results in a matching shift of $j - k$) or we need to find a smallest

```
Initialize
    1 ≤ j ≤ m: m_shift(j) := m;
    kmp_shift(m) := 1;
Stage 1
    j := m;
    do k = m - 1 to 1 by -1
        while p(j) ≠ p(k) do
            m_shift(j) := min{m_shift(j), j - k};
            if j < m then j := j + kmp_shift(j + 1)
                     else exit the while loop
            end;
        if p(j) = p(k)
            then do kmp_shift(k) := j - k; j := j - 1 end
            else (the loop was exited and so j = m)
                kmp_shift(k) := j - k + 1
        end;
Stage 2
    j := j + 1; j_old := 1;
    while j ≤ m do
        for i = j_old to j - 1 do m_shift(i) := min{m_shift(i), j - 1} end;
        j_old := j; j := j + kmp_shift(j)
        end
```

The program can be optimized, but at some loss in clarity.

Figure 13: Program to Compute the Matching Shift

$j > k$ such that $p[j,m]$ is a prefix of p (this results in a matching shift of $j-1$). These are computed in Stages 1 and 2, respectively.

Let us consider the computation of Stage 1. Basically, it seeks to find longest suffixes, $p[j,m]$, of the pattern that are proper prefixes of $p[k,m]$ (so $j > k$), for each k, $m > k \geq 1$. The algorithm proceeds by traversing two instances of the pattern, one indexed by j and one by k. The invariant is that $p[j+1,m]$ is a prefix of $p[k+1,m]$. If a match is found ($p(j) = p(k)$) we move one position to the left in both patterns; if no match is found, we slide the pattern indexed by j to the left by the minimum possible distance (i.e., so that the longest proper suffix of $p[j+1,m]$ which matches a prefix of $p[j+1,m]$ is now aligned with a prefix of $p[k+1,m]$; this amounts to the assignment $j := j + kmp\_shift(j+1)$, unless $j = m$, in which case we slide the pattern one cell to the left, which leaves j unchanged and decrements k). How could we fail to compute $m\_shift(j') = j' - k'$ for some j'? There are two possibilities. Either k is decremented to a value smaller than k' while $j < j'$, or j is incremented to a value that ensures the cell eventually compared to $p(k')$ is to the right of $p(j')$. The first possibility requires a match of $p(h)$ and $p(k')$ to be found for some $k' < h < j'$. But then $m\_shift(j') \leq j' - h$, contrary to assumption. The second possibility requires there to be h and l, $k' < h < j'$ and $l \geq 0$, with $p(k'+l)$ and $p(h+l)$ being matched, and $kmp\_shift(h+l+1) > j' - h$; but this also contradicts the fact that $m\_shift(j') = j' - k'$. So, in fact, Stage 1 correctly computes all the matching shifts of the first type.

Let us turn to Stage 2. In order to find $m\_shift(i)$, we need to know the least $j > i$ such that $p[j, m]$ is a prefix of p. But this is exactly what is computed by Stage 2.

It is straightforward to observe that the shift functions can be computed in time $O(m)$.

References

[AC75] A.V. Aho and M.J. Corasick. Efficient string matching: An aid to bibliographic search. *CACM*, 18(6), 333-340.

[AG86] A. Apostolico and R. Giancarlo. The Boyer-Moore-Galil string searching strategies revisited. *SIAM Journal on Computing*, Vol.15, 1(1986), 98-105.

[AHU73] A.V. Aho, J.E. Hopcroft, J.D. Ullman. *The design and analysis of algorithms*. Addison Wesley. 1973.

[ALV90] A. Amir, G.M. Landau, U. Vishkin. Efficient pattern matching with scaling. In *Proceedings of the First Annual ACM-SIAM Symposium on Discrete Algorithms*, 1990, 344-457.

[BB88] G. Brassard and P. Bratley. *Algorithmics. Theory & Practice*. Prentice Hall. 1988.

[BGR90] R. Baeza-Yates, G.H. Gonnet, M. Regnier. Analysis of Boyer-Moore type string searching algorithms. In *Proceedings of the First Annual ACM-SIAM Symposium on Discrete Algorithms*, 1990, 328-343.

[BM77] R. Boyer and S. Moore. A fast string matching algorithm. *CACM*, 20(1977), 762-772.

[CP89] M. Crochemore and D. Perrin. Two-way pattern matching. Technical Report, Laboratoire Informatique, Théorique et Programmation, Université Paris 7, 1989.

[Ga79] Z. Galil. On Improving the worst case running time of the Boyer-Moore string matching algorithm. *CACM*, 22(1979), 505-508.

[Ga81] Z. Galil. String matching in real time. *JACM*, Vol. 28 1(1981), 134-149.

[GP89] Z. Galil and K. Park. An improved algorithm for approximate string matching. In *Proceedings of the Sixteenth International Colloquium on Automata, Languages and Programming* 1989, 394-404.

[GS83] Z. Galil and J. Seiferas. Time space optimal string matching. *Journal Comput. Syst. Sci.* 26(1983), 280-294.

[GO80] L.J. Guibas and A.M. Odlyzko. A new proof of the linearity of the Boyer-Moore string searching algorithm. *SIAM Journal on Computing*, Vol.9, 4(1980), 672-682.

[KMP77] D.E. Knuth, J. Morris, V. Pratt. Fast pattern matching in strings. *SIAM Journal on Computing*, 6(1973), 323-350.

[KR87] R. Karp and M.O. Rabin. Efficient randomized pattern matching algorithms. *IBM Journal of Research and Development*, 31(2), 249-260.

[LV89] G.M. Landau and U. Vishkin. Fast parallel and serial approximate string matching. *Journal of Algorithms*, Vol.10 2(1989), 262-272.

[Ry80] W. Rytter. A correct preprocessing algorithm for Boyer-Moore string searching. *SIAM Journal on Computing*, Vol.9, 3(1980), 509-512.

[Uk85] E. Ukkonen. Finding approximate patterns in strings. *Journal of Algorithms*, Vol.6, 1985, 132-137.

[Vi85] U. Vishkin. Optimal pattern matching in strings. *Information and Control*, Vol.67, 1985, 91-113.

[Vi90] U. Vishkin. Deterministic sampling – A new technique for fast pattern matching. In *Proceedings of the Twenty Second Annual ACM Symposium on Theory of Computing*, 1990, 170-180.

Spotted something that isn't quite right?

Let us know and we'll fix it for future reprints. And as a thank you for your diligence, we'll send you a free Sothis book of your choice.

publisher@sothis.press

This page intentionally left blank

Except of course for that line above. And this one.
And whilst we have your attention, perhaps you'd
like to follow us on Instagram @sothis.press. Go on.

Made in United States
Orlando, FL
04 November 2026

90201491R00018